I0489251

ISBN: Softcover   978-1-4990-7544-1
        EBook   978-1-4990-7545-8

Rev. date: 09/27/2014

To order additional copies of this book, contact:
Xlibris LLC
1-888-795-4274
www.Xlibris.com
Orders@Xlibris.com

1 Isa
One

2 Dalawa
Two

Dalawang Pusa
Two cats

3 Tatlo
Three

Tatlong baboy
Three pigs

4 Apat
Four

Apat na palaka
Four frogs

5 Lima
Five

Limang Manok
Five chicken

6

Anim na pato
Six ducks

7 Pito
Seven

Pitong baka
Seven cows

8 Walo
Eight

Walong kambing
Eight goats

9 Siyam
Nine

Siyam na kabayo
Nine horses

10 Sampu
Ten

11

11 Labing-isa
Eleven

Labing-isang aso
Eleven dogs

12 Labindalwa
Twelve

Labindalwang pusa
Twelve cats

13 Labintatlo
Thirteen

Labintatlong baboy
Thirteen pigs

14 Labing-apat
Fourteen

Labing-apat na palaka
Fourteen frogs

15 Labinlima Fifteen

Labinlimang manok
Fifteen chicken

16 Labing-anim
Sixteen

Labing-anim na pato
Sixteen ducks

17 Labimpito Seventeen

Labimpitong baka
Seventeen cows

18 Labing-walo
Eighteen

Labing-walong kambing
Eighteen goats

19 Labinsiyam
Nineteen

Labinsiyam na kabayo
Nineteen horses

20 Dalawampu
Twenty

Dalawampung tupa
Twenty Sheep

Counting by Tens

10	Sampu
20	Dalawampu
30	Tatlumpu
40	Apatnapu
50	Limampu
60	Animnapu
70	Pitumpu
80	Walumpu
90	Siyamnapu
100	Isang Daan
1000	Isang Libo

More ways to count

Dalawampu't isa
Tatlumpu't dalwa
Apatnapu't tatlo
Limampu't apat
Animnapu't lima
Pitumpu't anim
Walumpu't pito
Siyamnapu't walo

Note: some words like dalawa is spelled dalwa when used in labindalwa, dalawampu't dalwa etc. for better pronunciation.